I0505598

Science with a Beat

THE A-B-C AMPHIBIAN AND REPTILE BOOK

WRITTEN AND ILLUSTRATED BY JACQUIE HAWKINS

PART OF THE
A-B-C SCIENCE SERIES

A

An **ALLIGATOR** is a reptile with roots from long ago.
It is known to live to be twenty-five years old.
It is sometimes dangerous and just may be a threat.
So I don't think a baby alligator's a good pet.
Fish, turtles, birds and crabs are what they like to eat.
Females lay a lot of eggs that get hatched from the heat.

The **ALLIGATOR LIZARD** is dull colored and lives alone.
It does well in captivity as a pet inside your home.
It grows up to ten inches long but is shy and will retreat.
Insects, spiders and the like are what it wants to eat.

2.

An
**ALLIGATOR
SNAPPER**
you don't
<u>want</u> hanging
<u>around</u>.
It <u>grows</u> to
thirty <u>inches</u>,
weighing one
<u>hundred</u> and
fifty <u>pounds</u>!

It lies <u>on</u> the muddy <u>bottom</u> and <u>tries</u> to catch a <u>fish</u>
By <u>waving</u> its wormlike <u>tongue</u> around.
The <u>fish</u> cannot <u>resist</u>.

A's <u>a</u>lso for
AM<u>PHI</u>BIANS.
They do <u>not</u> have
scaly <u>skin</u>
Like <u>reptiles</u>.
And they <u>don't</u>
have claws to
<u>use</u> to help them
w<u>in</u>.
They <u>mostly</u> live in
<u>water</u> or <u>places</u>
that are <u>wet</u>.
Some <u>kids</u> keep
frogs and <u>toads</u>
and sala<u>manders</u>
as a <u>pet</u>.

The **ANOLE** is
another name
for the
American
Chameleon.
It can change
its color and
that sounds
to me a
lot of fun.
If it's on bark it
turns to brown;
on a leaf it turns
to green.
When it changes color it is harder to be seen.

B

The
**BLANDING
TURTLE**
lives in
water
in a marshy
domain.
It is very shy

and so it's easier to tame.
It will make a good pet if to have one is your plan.
Just keep it in a little water in a shallow pan.

4.

The **BLIND SALAMANDER**
is unusual
indeed.
You'll find one in
a deep well or an
underground
stream.
Its eyes are not
developed.
There really is no
need

Because in dark, damp places is
where this guy will feed.

BOX TURTLES live
on land in open
woods or mucky
swamps.
Insects, berries,
worms and snails are
what they like to
chomp.
If you keep one as a
pet, it eats meat, fruit
and greens.

Keep him in an outdoor pen. It's easier than it seems.
If you'd rather keep him indoors,
there inside your house
You can, but it is better if you let him roam about.

A
BULLFROG
is a <u>large</u>
frog and it
<u>has</u> two real
large <u>ears</u>
That <u>lie</u>
behind its
<u>eyes</u>.

I think it <u>looks</u> a little <u>weird</u>.
The <u>color's</u> mostly <u>green</u>, a green that's
<u>very</u> drab and <u>drear</u>.
In <u>order</u> to ma<u>ture</u>, it takes a <u>tadpole</u> two long <u>years</u>.

C

The **CAVE
SALA<u>MA</u>NDER**
is <u>covered</u> up
with <u>spots</u>
On <u>yellow</u> or on
<u>orange</u>, just
<u>like</u> wild
polka-<u>dots</u>.
You'll <u>find</u> one
near a <u>cave</u>…
(I'm <u>sure</u> that's

not a <u>shock</u>.)
<u>Near</u> the entrance <u>to</u> a cave's <u>overhanging</u> <u>rock</u>.

6.

The **CHORUS FROG** or **CRICKET FROG** really doesn't sing.
It's also called a Tree Frog but it doesn't climb a thing.
Some might climb up to an inch but most don't climb at all.

It measures less than two inches and so are pretty small.

COMMON SNAPPERS are dangerous for powerful are their jaws.
They have a vicious temper so don't handle them at all.
They prefer the quiet, muddy water of a pond.
Waterfowl and fish is what they like to feed upon.

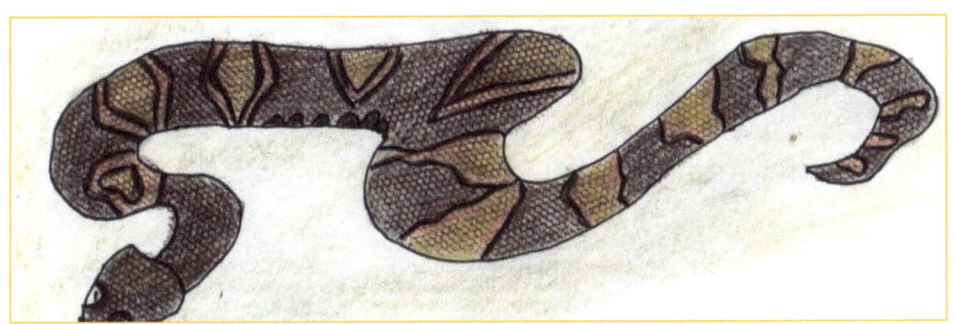

A **COPPERHEAD** is <u>poisonous</u>, much <u>like</u> a rattle<u>snake</u>.
The <u>pits</u> between its <u>eyes</u> and nostrils
<u>tells</u> him where to <u>strike</u>
And <u>he</u> might just strike <u>out</u> at you if
<u>you</u> should try to <u>pass</u>.
So, <u>if</u> you come <u>across</u> one better <u>get</u> away real <u>fast</u>!.

The **COTTON MOUTH**, or <u>water</u> moccasin, <u>is</u> a snake more <u>vicious</u>! It <u>feeds</u> on fish and <u>frogs</u> because

to <u>it</u> they are <u>delicious</u>.

8.

The **CORAL** **SNAKE'S**
another one that you
should surely dread,
Though it's pretty with its
rings of black, yellow
and red.
It is highly poisonous
and I have heard it said
That if one ever
bites you,
well you just might
end up dead!
There are other snakes

made up of
black and red and yellow.
But "if red touches yellow then he's not a friendly fellow."

D

The **DUSKY**
SALAMANDER is
difficult to be seen.
Its mottled skin blends
in with rocks and
moss along a stream.

The **DESERT IGUANA**
or Crested Lizard's a
handsome little 'feller'.
It burrows under
shrubs and is an open
desert
dweller.
Almost twice as long as
his whole body is his
tail.
He runs so fast he's
hard to catch as over
the sand he sails.

E

Perhaps the best
of salamanders
kids keep as
a pet
Is the prettiest
and most
interesting one
they called
RED EFT.
It does well in terrariums and eats small bits of meat.
It lives on land but returns to water
when it wants to breed.

10.

F

FADED SNAKES
don't really fade
but I
guess they move
along.
They constrict and
squeeze their prey
with bodies that are
strong.

**FOOTLESS
LIZARDS** look like
worms because
they have no
limbs.
But I guess no
feet or legs is
quite alright
with them.
Because they like
to burrow, legs
would just get in
their way.

They're found only in Florida or California.

A **FOX SNAKE** does
not <u>look</u> much like a
<u>fox</u> as its name
<u>claims</u>.
S**o** I <u>don't</u> know how it
<u>ever</u> got the <u>fox</u>
put in its <u>name</u>...
Though <u>like</u> a fox it
<u>lives</u> outdoors, its
<u>habitat's</u> the <u>same</u>.
Un<u>like</u> a fox this
<u>snake</u> can very <u>easily</u>
be <u>tamed</u>.

The **FOUR-TOED
SALAMANDER'S**
name
<u>is</u> I would <u>suppose</u>
Be<u>cause</u> on all of
<u>its</u> four feet
<u>there</u> are just four
<u>toes</u>.
<u>You</u> will find it <u>in</u> a
swamp or <u>in</u> thick
wooded <u>thatch</u>.
The <u>mother</u> lays

her <u>eggs</u>, then stays <u>until</u> they all have <u>hatched</u>.

12.

G

**GARTER
SNAKES** are <u>very</u>
common
throug<u>hout</u> the
U.S.<u>A.</u>;
<u>Ribbon</u> Garters <u>in</u>
the east;

<u>others</u> live in the <u>western</u> states.
Their <u>young</u> are born <u>alive</u> in summer, <u>twenty</u> at a <u>time</u>!
<u>They</u> choose frogs or <u>toads</u> or earthworms
<u>when</u>ever they <u>dine</u>.

GECKOS are
so <u>very</u> cute!
They <u>get</u> about
with <u>ease</u>.
<u>They</u> have
padded <u>toes</u>
and lidless <u>eyes</u>
and live near
<u>homes</u> and
trees.
<u>Geckos</u> are
quite <u>docile</u> and
they <u>very</u> rarely
<u>bite</u>.
<u>Geckos</u> are noc<u>turnal</u> so they <u>come</u> out late at <u>night</u>.

Though the
GILA MONSTER
has a really
monstrous bite
It's not exactly a
monster, but I
guess that's
being trite.
It's the only
poisonous lizard
with an attitude
to fight
And when it bites
it won't let go but hangs on for dear life.

**GIANT
SALAMANDERS**
are very large
indeed....
Some of them a
whole foot long,
the largest of land
species.
They're found on
moist slopes,
under rocks
and logs
or near a stream.
Mottled black and
dark legs

seem to be its color scheme.

14.

GLASS SNAKE LIZARDS do not look like lizards….not at all! They have no legs or feet, so like a snake… they do not crawl. If they're roughly handled it could really be a turn-off Because like glass

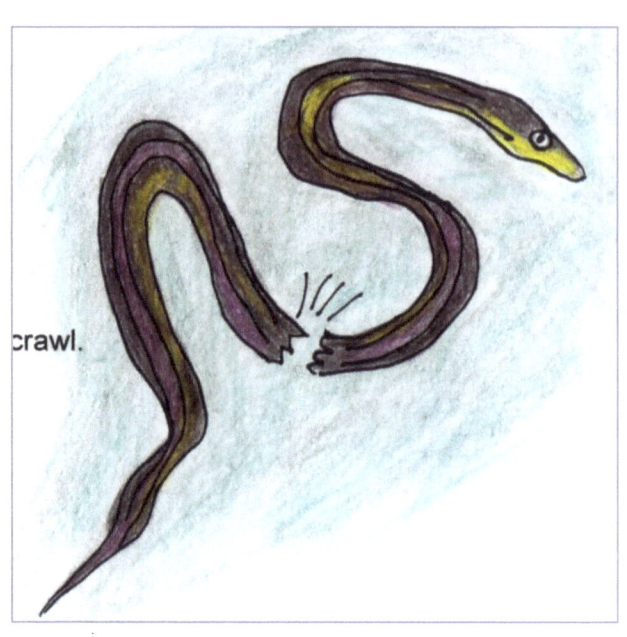

crawl.

they break in two and a part will simply fall-off.

The **GOPHER FROG** gets not its name because of how it looks

But because it steals a gopher turtle's home…
much like a crook.
It is grayish green and has a bunch of small black spots.
It's really small so I don't know why it acts like a big-shot!

Tortoises or **GOPHER TURTLES** have blunt, club-like feet.
They do not have webbed feet so they are not aquatic species.
It looks like tiny pyramids or mountains on their backs.
I guess it wards off predators who might want to attack.

H

THE HELL-BENDERS are aquatic.
They have a wrinkled skin.
Under rocks in shallow water's where they like to swim.
Some are spotted yellowish; some are red or brown.
The female lays a mass of eggs in the muddy ground.

16.

The **HOG-NOSED SNAKE'S** <u>turned</u>-up nose
<u>is</u> what helps it <u>dig</u>.....
For <u>yummy</u> toads.
It <u>eats</u> a lot be<u>cause</u> it is quite <u>big</u>.
<u>If</u> attacked it <u>hisses</u>, strikes, puffs <u>up</u>, acts mean <u>although</u>…
<u>He</u> is actually harmless for it <u>never</u> actually <u>bites</u>.
<u>If</u> that doesn't <u>work</u> he just rolls <u>over</u> and plays <u>dead</u>.
It <u>is</u> a gentle <u>snake</u> and its <u>habitat's</u> wide<u>spread</u>.

The **HYLA** is a <u>tree</u> frog and some <u>barely</u> are an <u>inch</u>.
<u>That</u> it is not <u>danger</u>ous to <u>humans</u> is a <u>cinch</u>.
It <u>clings</u> to leaves and <u>twigs</u> and such with <u>very</u> sticky <u>toes</u>.
Its <u>call</u> is clear and <u>musical</u>. It <u>sings</u> wherever it <u>goes</u>.
Its <u>colors</u> and its <u>patterns</u> are <u>rarely</u> just the <u>same</u>
And <u>like</u> Chameleons, <u>if</u> they want, their <u>color</u> they can <u>change</u>.

HORNED LIZARDS
are <u>unique</u> and they
are <u>only</u> found out
<u>West</u>.
<u>Some</u> puff up or
<u>flatten</u> out when<u>ever</u>
they're <u>distressed</u>.
<u>Some</u> may even
<u>squirt</u> blood from the
<u>corner</u> of their
<u>eyes</u>....
To <u>chase</u> away a <u>foe</u>.
It's just <u>part</u> of their <u>disguise</u>.
<u>They</u> are easily <u>captured</u> and <u>safely</u> become <u>tame</u>,
So <u>acting</u> mean and <u>ugly</u> is just <u>all</u> part of their <u>game</u>.

**INDIGO
SNAKES**
<u>are</u> among
the <u>largest</u>
of our
<u>snakes</u>.
<u>They</u> can
be up to
<u>eight</u> feet
long! That's <u>long</u> for pity's <u>sake</u>!
They're <u>heavy</u> and they're <u>handsome</u>…a
<u>shiny</u> midnight-<u>blue</u>.
They're <u>easily</u> tamed when <u>captured</u> and can
<u>be</u> a friend to <u>you</u>.

18.

J

J is for the **JEFFERSON SALAMANDER**
found up North…
Where it's cold in Maine and all
along the Great Lake ports.
It's sometimes called Blue-
Spotted for the blue spots
on its skin.
It likes the woods and swamps and streams so in
water it can swim.

California King Snake

Red King Snake

Common King Snake

Scarlet King Snake

There are all kinds
of **KING SNAKES**
found in almost
every state
All across the U.S.
and they surely do
look great.
Some of them are
so bright they'd be
easy to locate
And their patterns
are quite different
on different kinds of snakes.
The Scarlet King is like the Coral with
red and black and yellow.
But the yellow doesn't touch the red so
he's a friendly fellow.

K

K is for the **KEY** to learning <u>all</u> about these <u>creatures</u>.
<u>Books</u> can tell you <u>all</u> about them <u>and</u> include each <u>feature</u>
So <u>if</u> this book of <u>rhymes</u> is not all <u>that</u> you're looking <u>for</u>…
Just <u>head</u> out to your <u>library</u> for <u>you</u> will find lots <u>more</u>.

L

L is for the **LEOPARD LIZARD** <u>named</u> for all its <u>spots</u>.
It <u>likes</u> flat, sandy <u>deserts</u> or it <u>hides</u> among the <u>rocks</u>.
It <u>tries</u> to eat small <u>insects</u> and <u>lizards</u> that it <u>finds</u>

But <u>if</u> it cannot <u>find</u> some, why he'll <u>just</u> eat his own <u>kind</u>.
This <u>kind</u> of lizard's <u>vicious</u> and it <u>won't</u> make a good <u>pet</u>.
So <u>if</u> you want a <u>lizard</u>…the <u>leopard</u> one don't <u>get</u>.

20.

The **LONG-TAILED SALAMANDER** is found only in the East. Its tail is longer than its body, which is quite a feat. Its skin is orange or yellow with black

splotches here and there.
It's a real cute little fellow with a lot of flair.

I don't know why the **LONG-NOSED SNAKE'S** name somebody chose Because you see it really does not have a real long nose.

Like any snake that slithers this one is both foot and armless
But unlike so many this one is completely harmless.

M

**MARBLED
SALA-
MANDERS**
do not
look

marbled at <u>all</u>.
They're <u>just</u> four inches <u>long</u> and so they're
<u>really</u> kind of <u>small</u>.
<u>They</u> are kind of <u>stout</u> with light gray
<u>markings</u> on black <u>skin</u>.
I <u>think</u> they're kind of <u>cute</u> if you <u>look</u> at them close-<u>in</u>.

**MYHLENBERG
TURTLES**,
<u>usually</u> live
on <u>land</u>
But <u>go</u> back to
the <u>water</u> when
<u>danger</u> is
at <u>hand</u>.
<u>They</u> have orange <u>blotches</u> on <u>both</u> sides of their <u>heads.</u>
So <u>you</u> can recog<u>nize</u> them, or <u>so</u> it has been <u>said</u>.
<u>They</u> make a nice <u>pet</u> because they're <u>real</u> easy to <u>feed</u>…
<u>Some</u> greens, fruit and <u>earthworms</u>
<u>along</u> with some chopped <u>meat</u>.

22.

MUSK TURLES,
sunning
themselves,
is what's
most often seen
In ponds and in
slow moving
rivers
or slow moving
streams.
They have a
musky odor,
which gave to
them

their name.
That it wouldn't make a good pet should be really plain.

N

**NARROW-
MOUTHED
FROGS**
of the East
are very
dark.
The
western
ones are

light with hardly any kind of mark.
They have tiny voices and come out late at night.
They hide under rocks and logs when
scared and take to flight.

NEWTS <u>never</u> do lay <u>eggs</u>, instead their
<u>young</u> are born <u>alive</u>.
<u>They</u> do not have <u>any</u> eyelids <u>at</u> all on their <u>eyes</u>.
They <u>are</u> nocturnal….<u>hunt</u> at night <u>when</u> the sky is <u>black</u>
<u>I</u> guess when the <u>sun</u> comes up they <u>hop</u> into the <u>sack</u>.

The **NIGHT SNAKE** is quite <u>poisonous</u> but <u>it</u> does not have <u>fangs</u>.

<u>If</u> it bites you, <u>its</u> saliva's <u>what</u> gives all the <u>pain</u>.
<u>So</u> a '**Fangless**' <u>Night</u> Snake is <u>actually</u> its <u>name</u>.
<u>It</u> bites little <u>lizards</u> and <u>that</u> is how they're <u>slain</u>.

24.

O

OLYMPIC SALA- MANDERS are <u>really</u> not <u>athletes</u>. They <u>do</u> not train to get <u>strong</u> and I <u>don't</u>

think they <u>compete</u>.
<u>They</u> live in the <u>forest</u> on the <u>California</u> <u>coast</u>…
<u>Beneath</u> the logs and <u>rocks</u> because they
<u>like</u> things damp the <u>most</u>.

P

PACIFIC TURTLES don't <u>live</u> in oceans though they're <u>found</u> that far out <u>west</u>. They're the <u>only</u> turtles <u>out</u> there that <u>like</u> their water <u>fresh</u>. <u>They</u> prefer a
<u>mountain</u> lake or a <u>stream</u> with slower <u>flow</u>.
It's <u>said</u> that they can <u>make</u> good pets
but I <u>really</u> wouldn't <u>know</u>.

The **PAINTED SALA-MANDER** is
another western species.
You could try to paint one
but I guess it's not that easy.
It's kind of cute with blotches orange, yellow or some red
Alternating with black blotches
that throughout are spread.
It lives in mountain oak trees or in forest evergreens.
It has a complex courtship and to
watch one is a scream.

PATCH-NOSED SNAKES are
very fast.
They must be
pretty light.
You can recognize
them by their
pretty yellow stripe.

26.

The **PURPLE
SALAMANDER**
is not purple!
Not at all!
It's brown to
reddish-brown
with splotches
dark; throughout
they're sprawled.
It lives on hills or
in the mountains.
By streams or ponds
it lays
Up in the Appalachian
Mountains
in the Northeast U.S.A.

Q

Q is for the
QUESTIONS that
you still might
want to ask
About reptiles
and amphibians. You might want the answers fast.
Using a computer is a good way to get facts.
Type in a question and real soon
the answer you'll get back!

R

The **RAINBOW SNAKE** is <u>not</u> a rainbow. <u>There</u> is little <u>doubt</u>. <u>But</u> the colors <u>on</u> the top and <u>bottom</u> turn-<u>about</u>.

The <u>top</u> part of it's <u>black</u> with two full <u>rows</u> of reddish <u>dots</u>
While the <u>bottom</u> is all <u>reddish</u> with two <u>full</u> rows of black <u>spots</u>!

RATTLESNAKES have <u>rattles</u> although <u>I</u> would be <u>afraid</u> To <u>ever</u> try to <u>give</u> this rattle <u>to</u> a tiny <u>babe</u>! The <u>Diamond</u>back is <u>famous</u> for its <u>pretty</u> diamond <u>shapes</u>. <u>It's</u> the largest <u>kind</u>; The record <u>is</u> a nine foot <u>snake</u>! You <u>wouldn't</u> want to <u>own</u> one for it <u>is</u> a poisonous <u>one</u>.

<u>If</u> you hear it <u>rattle</u>………it's a <u>warning</u>….better <u>run</u>!

28.

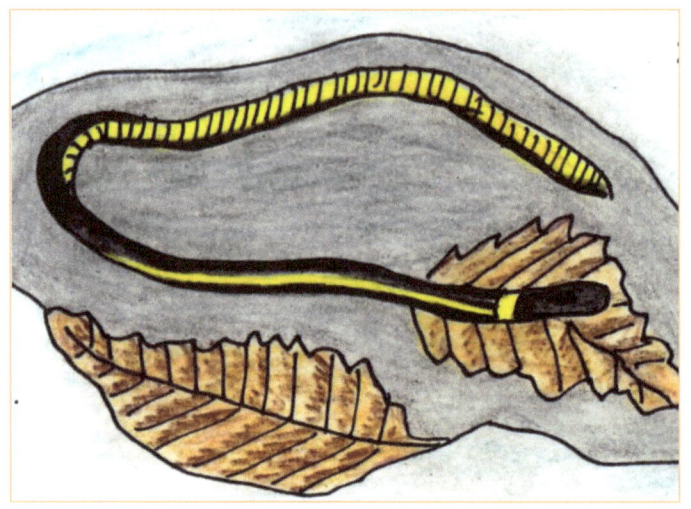

The **RING-NECKED SNAKE**
is slate gray on
the top but
then below
It's yellow, red
or orange
which is usually
what will show.
The ring
around its neck is like a necklace made of gold.
That's what you should look for so
this snake you will then know.

The **ROBBER FROG** is
only found in places
where it's hot
Like Florida and Texas.
There are really not a lot.
It is short;
it is squat
and has a real fat head.
It sounds like a
barking dog,
at least that's
what is said.
Because this tiny frog
sometimes will

sound like a pet dog
People sometimes call it by the name 'The Barking Frog'.

S

The **SAND SNAKE** is a burrower that lives in desert sands.
Its color's red to yellow interspersed with real dark bands.

SKINKS live almost anywhere in the U.S.A.
No other lizard can be found with so great a range.
They're about the only lizard folks can see up north.
Most of them run swiftly though their legs are really short.

S<u>OFT</u> SHELLED TURTLES <u>have</u> hard shells
though <u>soft</u> along the <u>rim</u>.
They <u>have</u> long necks and <u>sharp</u> beaks and are
<u>vicious</u> on a <u>whim</u>.
<u>They</u> are pretty <u>heavy</u> and can <u>weight</u> to thirty <u>pounds</u>.
You <u>wouldn't</u> want one <u>as</u> a pet or <u>even</u> hanging <u>around</u>.

SWIFTS are
really <u>hard</u>
to catch
be<u>cause</u> they
are so <u>fast</u>.
<u>If</u> they see you
<u>coming</u>.........
<u>away</u> they'll
quickly <u>dash</u>.
<u>So</u> to call them <u>Swifts</u> I guess is <u>really</u> a good <u>name</u>.
<u>If</u> this pet gets <u>loose</u> inside your <u>house</u>...
there's you to <u>blame</u>.

The **SPOTTED TURTLE** <u>is</u> named for the <u>spots</u> around its <u>shell</u>.
The <u>spots</u> are round and <u>make</u> it look like <u>dominoes</u> that <u>fell</u>.
It <u>makes</u> a real nice <u>long</u>-lived pet that <u>really</u> is quite <u>neat</u>.
Just <u>feed</u> it some fresh <u>lettuce</u>, some <u>fish</u> and bits of <u>meat</u>.

<u>SLIMY SALAMANDERS</u> lay their <u>eggs</u> on rotting <u>bark</u>.
<u>If</u> they're really <u>slimy</u> then I <u>think</u> they must be <u>smart</u>.
I <u>wouldn't</u> want to <u>pick</u> one up or <u>for</u> a pet one <u>own</u>.
<u>No</u>! I think that <u>I</u> would simply <u>leave</u> the thing <u>alone</u>!

T

The **TAILED TOAD** that lives <u>in</u> the Northwest <u>actually</u> has a <u>tail</u>!
<u>Though</u> to wag it <u>when</u> he's happy…is <u>just</u> a <u>fairy</u>tale.

TERRAPINS are <u>raised</u> on turtle <u>farms</u> for their good <u>meat</u>. It <u>is</u> the best known <u>turtle</u> and to <u>eat</u> one is a <u>treat</u>. In <u>many</u> states young <u>Terrapins</u> are <u>protected</u> by the <u>law</u>. <u>Adult</u> ones sell for <u>lots</u> of money <u>even</u> though they're <u>raw</u>.

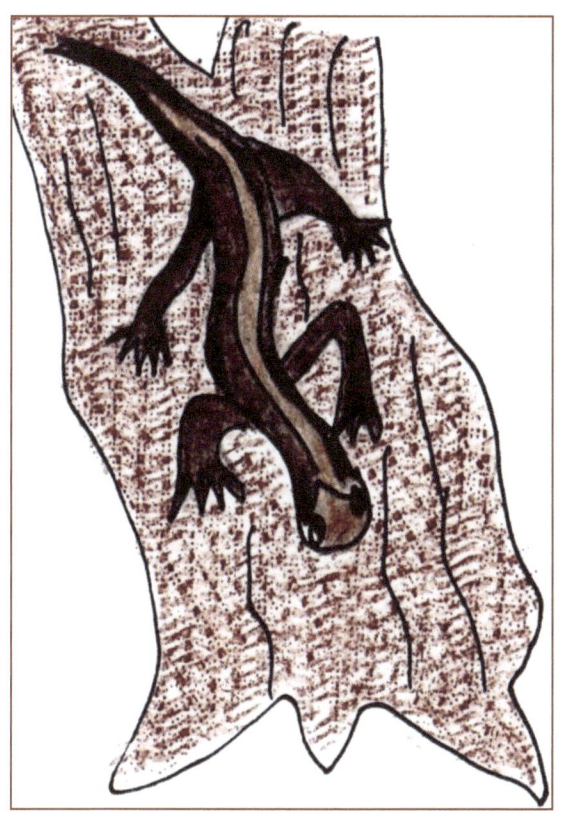

The <u>Pacific</u> Coast
TREE <u>SALAMANDER</u>
<u>lives</u> inside of <u>trees</u>
In <u>little</u> cav<u>ities</u> and
sometimes <u>in</u> a colo<u>ny</u>.
It <u>also</u> can live <u>under</u>
logs or <u>rocks</u> or
on the <u>ground</u>.
Its <u>color</u> is light <u>brown</u>
with little <u>markings</u>
to be <u>found</u>.

<u>You</u> can only <u>see</u>
a **TRUE <u>IGUANA</u>**
in a <u>zoo</u>
Be<u>cause</u> it lives in
<u>Mexico</u> and
<u>Central</u> America
<u>too</u>.
It's <u>four</u> to six feet
<u>long</u> and lives up
<u>in</u> the tropical
<u>trees</u>.

It <u>also</u> lives on the <u>Galapagos</u> Islands
<u>and</u> in the West <u>Indies</u>.

U

Some **UTAS** live on <u>rocks</u> and others <u>live</u> up in the <u>trees</u>. <u>Some</u> are striped or <u>speckled</u> and on <u>small</u> insects they <u>feed</u>. <u>If</u> you want to <u>catch</u> one it is

<u>best</u> to do at <u>night</u>
Be<u>cause</u> they are less <u>active</u> and won't
<u>scurry</u> out of <u>sight</u>.

V

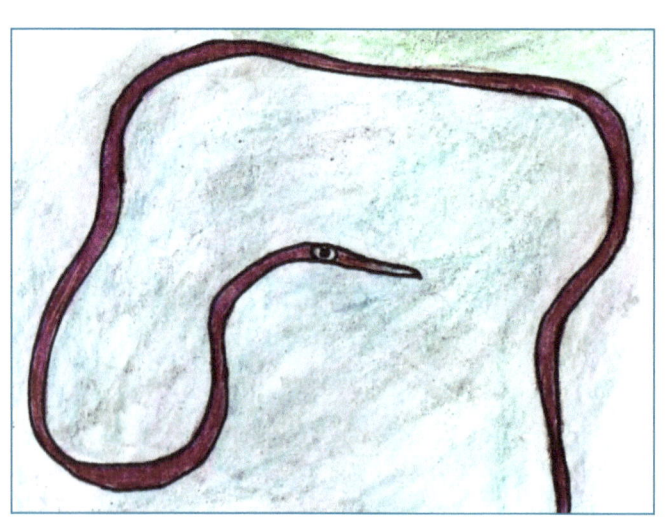

<u>VINE</u> **SNAKES** **a**re brush <u>dwellers</u>, four feet <u>long</u> or maybe <u>more</u>. <u>They</u> are long and <u>skinny</u>. To <u>me</u> they look like <u>cords</u>.

W

A
WATERDOG'S
a <u>salamander</u>
that <u>likes</u> to live
in <u>lakes</u>
Or <u>rivers</u>….
For the <u>love</u> of
water <u>they</u> do
not for<u>sake</u>.

<u>It</u> looks really <u>weird</u> with its <u>bushy</u>, reddish <u>gills</u>.
<u>Breathing</u> under<u>water</u> like a <u>fish</u> must give him <u>thrills</u>!
I <u>don't</u> think that a <u>Waterdog</u> can <u>bark</u> or beg or <u>catch</u>,
And <u>if</u> you throw a <u>stick</u> for him I <u>doubt</u> it he would <u>fetch</u>.

WHIPSNAKES are
both <u>long</u> and thin,
<u>fast</u> and hard
to <u>grab</u>.
They <u>do</u> not make
good <u>pets</u> because
they're <u>really</u>
kind of <u>bad</u>.
When <u>they</u> are
caught they <u>bite</u> and
whip their <u>tails</u>
around real <u>fast</u>.
So <u>if</u> you think you
<u>want</u> one, on that
<u>plan</u> you'd better
<u>pass</u>.

36.
Did you know there
is a frog that whistles
while it works?
The **WHISTLING
FROG** can make a
faint, soft whistling
little chirp.
Its eggs don't turn to
tadpoles for the
eggs are laid on land
And when they hatch
they have   their legs
already and can
stand!

The **WOOD TURTLES**
I'm sure you know are
not made out of wood.
They got their name
because they like
the woods….just as
they should.
But they leave the
woods for open land
when they do feed.
When they get dry, to
swamps they go,
or ponds and
sometimes streams.
Their skin is bright red
orange and their
shell looks really grand.
They make good pets and even will eat
berries from your hand.

The **WORM SALAMANDER** actually looks like a long worm.....
Although it has four little legs.
So I guess it doesn't have to squirm.
It is very thin and has on it a real long tail.
You'll find its eggs tucked under rotting leaves left by the female.

X

If you have lots of questions I suggest you write them down
And then you can X them out when the answers you have found.
It's a great way to keep up with all you want to know.
Learning is a great way to help your brain to grow.

How many fox snakes exist today?
How deep do turtles swim?
How long does it take for venum to take affect?
Is a terrapin a tortise or a turtle?
Do any turtles bite?

Y

Did you know that **YELLOW-LIPPED SNAKES**
have a yellow lip?
I didn't know snakes
have lips!
No, I didn't know a bit.
If they use yellow lipstick I am
sure it is a hit.
Under logs and leaves is where they
usually want to slip.

Z

Z is for a **ZOO**. It's where a <u>lot</u> of these you'll <u>find</u>
<u>As</u> you **ZIP** and **ZOOM** around the <u>place</u>
through paths that <u>twist</u> and wind.
ZOOS have many <u>animals</u> and so <u>many</u> different <u>kinds</u>
That <u>you</u> will be <u>amazed</u>! It will <u>surely</u> stretch your <u>mind</u>!

There <u>are</u> a lot more <u>snakes</u> than these, plus
<u>more</u> amphib<u>ians</u>
And <u>there</u> are lots of <u>books</u> I know that
<u>you</u> will find them <u>in</u>.
But <u>these</u> are some to <u>learn</u> about,
<u>if</u> you want to <u>know</u>
<u>About</u> snakes, lizards, <u>salamanders</u>,
<u>turtles</u>, frogs and <u>toads</u>.

LOOK FOR OTHER BOOKS
FROM THE A-B-C SCIENCE SERIES:

THE ABC SEASHELL BOOK

THE ABC TREE BOOK

THE ABC DINOSAUR BOOK

THE ABC WILDFLOWER BOOK

THE ABC BIRD BOOK

THE ABC BUG BOOK

THE ABC BUTTERFLY BOOK

THE ABC ZOO BOOK

THE ABC BODY BOOK

www.ingramcontent.com/pod-product-compliance
Lightning Source LLC
Chambersburg PA
CBHW040929180526
45159CB00002BA/673